Sounds of Sustainability Pollution Prevention Hearables

Tayne Divillers

Copyright © [2023]

Title: Sounds of Sustainability Pollution Prevention Hearables
Author's: Tayne Divillers

This book was printed and published by [Publisher's: **Tayne Divillers**] in [2023]

ISBN:

TABLE OF CONTENT

Chapter 1: Introduction to Pollution Prevention Hearables   06

Understanding Environmental Protection

The Role of Technology in Environmental Sustainability

Introduction to Pollution Prevention Hearables

Chapter 2: The Basics of Pollution Prevention Hearables   12

What are Pollution Prevention Hearables?

How Pollution Prevention Hearables Function

Advantages and Limitations of Pollution Prevention Hearables

Chapter 3: Types of Pollution Prevention Hearables   18

Noise-Canceling Headphones

Air Quality Monitoring Earbuds

Water Pollution Detection Earpieces

Soil Contamination Prevention Devices

Chapter 4: Applications of Pollution Prevention Hearables 26

Noise Pollution Reduction

Air Pollution Monitoring and Protection

Water Pollution Prevention and Detection

Soil Contamination Prevention and Rehabilitation

Chapter 5: Technological Innovations in Pollution Prevention Hearables 24

Artificial Intelligence Integration

Internet of Things Connectivity

Advanced Sensor Technologies

Sustainable Materials in Hearable Devices

Chapter 6: Challenges and Opportunities in Environmental Protection Hearables 42

Legal and Regulatory Considerations

Ethical Implications of Hearable Technologies

Economic and Market Trends

Future Opportunities and Potential Development

Chapter 7: Case Studies on Pollution Prevention Hearables   50

Case Study 1: Noise Reduction Hearables in Urban Environments

Case Study 2: Air Quality Monitoring Hearables in Industrial Settings

Case Study 3: Water Pollution Detection Hearables in Marine Environments

Case Study 4: Soil Contamination Prevention Hearables in Agricultural Practices

Chapter 8: The Future of Pollution Prevention Hearables   58

Emerging Technologies and Innovations

Integrating Hearables with Sustainable Lifestyles

Potential Impact on Environmental Conservation Efforts

Chapter 9: Conclusion   64

Recap of Key Points

Final Thoughts on Pollution Prevention Hearables

Call to Action for a Sustainable Future

Chapter 1: Introduction to Pollution Prevention Hearables

Understanding Environmental Protection

In today's rapidly evolving world, it is imperative that we all play a role in protecting our environment. As the global population continues to grow, so does the strain on our natural resources and ecosystems. It is crucial that we comprehend the concept of environmental protection and the steps we can take to preserve our planet for future generations.

Environmental protection refers to the practice of conserving and safeguarding natural resources, ecosystems, and the overall environment. It encompasses a wide range of actions aimed at reducing pollution, preventing habitat destruction, and promoting sustainable practices. By understanding the importance of environmental protection, we can make informed choices and contribute to a healthier planet.

One aspect of environmental protection that often goes unnoticed is the impact of pollution on our surroundings. Pollution is a major threat to our environment, leading to the degradation of air, water, and soil quality. It affects not only the health of humans but also the well-being of countless plant and animal species. By recognizing the detrimental effects of pollution, we can take steps to reduce our carbon footprint and minimize our contribution to environmental degradation.

Another crucial component of environmental protection is the preservation of natural habitats. Ecosystems provide a multitude of benefits, including clean air, clean water, biodiversity, and climate regulation. Unfortunately, human activities such as deforestation, urbanization, and industrialization have led to the destruction of many habitats. By understanding the importance of preserving these ecosystems, we can support initiatives aimed at conservation, reforestation, and the protection of endangered species.

For those interested in the niche of hearable beauty, understanding environmental protection is especially relevant. The production of hearable devices often involves the use of precious resources, energy consumption, and waste generation. By embracing sustainable practices, such as using recycled materials and reducing energy consumption during manufacturing, the hearable beauty industry can contribute to environmental protection.

In conclusion, understanding environmental protection is crucial for everyone, including those interested in the niche of hearable beauty. By recognizing the impact of pollution and the importance of conserving natural habitats, we can make informed choices and contribute to a sustainable future. The concept of environmental protection should be an integral part of our daily lives, guiding our actions towards a healthier and more resilient planet. Let us all join hands in protecting our environment for the benefit of current and future generations.

The Role of Technology in Environmental Sustainability

In today's rapidly evolving world, technology plays a crucial role in shaping our society and the way we interact with the environment. From reducing pollution to conserving resources, innovative technological advancements have opened up new avenues for promoting environmental sustainability. This subchapter will explore how technology, particularly in the realm of hearables, can contribute to the preservation of our planet.

Hearables, a niche in the realm of technology, have gained immense popularity in recent years due to their multifunctional capabilities and seamless integration into our daily lives. These devices, ranging from wireless earbuds to smart headphones, not only enhance our audio experience but can also be utilized to promote environmental sustainability.

One significant aspect of hearable technology in relation to environmental sustainability is its potential in pollution prevention. Traditional headphones and earphones often rely on disposable batteries, contributing to electronic waste. However, with the advent of rechargeable and eco-friendly batteries, hearables have become more sustainable. By utilizing hearables with rechargeable batteries, individuals can reduce their carbon footprint and play a part in minimizing electronic waste.

Furthermore, hearable beauty is another niche that has emerged, combining fashion and technology to create aesthetically pleasing devices. By incorporating sustainable materials and manufacturing practices, companies can produce hearables that not only function flawlessly but also contribute positively to the environment. These eco-friendly hearables not only promote sustainable fashion but also serve as a reminder to users about the importance of environmental conservation.

Moreover, technology in hearables can assist in raising awareness about environmental issues. Through the integration of sensors and advanced algorithms, hearables can track and analyze personal environmental data, such as noise pollution levels or air quality. By providing users with real-time feedback about their surroundings, these devices can encourage individuals to make eco-conscious choices and take actions to mitigate their impact on the environment.

In conclusion, technology, particularly in the realm of hearables, has an integral role to play in environmental sustainability. From pollution prevention to raising awareness, these innovative devices offer a range of possibilities to contribute positively to our planet. By embracing eco-friendly hearables and utilizing their potential, we can all become active participants in safeguarding the environment for future generations.

Introduction to Pollution Prevention Hearables

In the fast-paced world we live in, technology has become an integral part of our daily lives. From smartphones to smartwatches, we have witnessed the evolution of wearable devices that enhance our productivity and connectivity. However, in the pursuit of innovation, we often overlook the environmental impact of these gadgets. That's where pollution prevention hearables come into play.

This subchapter introduces you to the fascinating world of pollution prevention hearables and aims to raise awareness about their importance in promoting sustainable living. While the term "hearables" primarily refers to devices worn on or around the ears for audio purposes, pollution prevention hearables go a step further by incorporating eco-friendly practices and materials.

For the audience interested in hearable beauty, pollution prevention hearables offer a unique blend of style, functionality, and environmental consciousness. These devices not only provide a seamless audio experience but also prioritize sustainability, making them a perfect fit for those who value both aesthetics and the planet.

One of the key aspects of pollution prevention hearables is their focus on reducing electronic waste. With the increasing demand for technology, electronic waste has become a major environmental concern. Pollution prevention hearables address this issue by incorporating modular designs and sustainable materials that can be easily repaired, upgraded, or recycled. This not only extends the lifespan of the devices but also minimizes the need for constant replacements, reducing the overall waste generated.

Moreover, pollution prevention hearables strive to be energy-efficient. By optimizing power management systems and utilizing renewable energy sources, these devices aim to reduce energy consumption and minimize their carbon footprint. This commitment to sustainability ensures that you can enjoy your favorite music or podcasts without compromising the health of our planet.

In conclusion, pollution prevention hearables offer a refreshing approach to the world of wearable technology. By combining style, functionality, and environmental consciousness, these devices cater to the needs of the audience interested in hearable beauty while promoting sustainable living. Embracing pollution prevention hearables not only enhances our audio experience but also contributes to a healthier and greener environment for all.

Chapter 2: The Basics of Pollution Prevention Hearables

What are Pollution Prevention Hearables?

In today's world, where environmental issues and sustainability have taken center stage, a new concept has emerged to tackle pollution in a unique and innovative way - Pollution Prevention Hearables. These groundbreaking devices combine the realms of technology and fashion to create a solution that not only protects our planet but also enhances our personal style. This subchapter explores the concept of Pollution Prevention Hearables, their significance in the context of sustainability, and their appeal to the niche of "hearable beauty."

Pollution Prevention Hearables are wearable devices that serve multiple functions while actively working to reduce pollution. These devices are designed to be worn as fashionable accessories, such as earrings, necklaces, or bracelets, and are embedded with cutting-edge technology to monitor and mitigate pollution levels in our surroundings. They utilize sensors and advanced algorithms to detect harmful pollutants in the air, water, or soil, providing real-time data and alerts to the wearer. This enables individuals to make informed decisions about their surroundings and take necessary actions to minimize exposure to pollutants.

The significance of Pollution Prevention Hearables lies in their ability to empower individuals to actively participate in the fight against pollution. By wearing these devices, people become more aware of the environmental impact of their actions and surroundings. The data collected by Pollution Prevention Hearables can be shared with local authorities and environmental organizations, contributing to a collective effort in identifying pollution sources and implementing effective pollution prevention strategies.

One particular niche that Pollution Prevention Hearables cater to is "hearable beauty." These devices not only serve a practical purpose but are also designed with elegance and style in mind. From intricately crafted earrings to sleek bracelets, Pollution Prevention Hearables offer a wide range of options that seamlessly blend with any fashion statement. By incorporating these devices into their daily lives, individuals can make a fashion statement while actively contributing to a sustainable future.

In conclusion, Pollution Prevention Hearables are a remarkable innovation that combines technology, fashion, and sustainability. These devices allow individuals to monitor pollution levels, make informed decisions, and actively participate in the fight against environmental degradation. The niche of "hearable beauty" further adds to their appeal by offering stylish and elegant options that align with personal fashion choices. By embracing Pollution Prevention Hearables, we can pave the way for a cleaner, healthier, and more sustainable planet for ourselves and future generations.

How Pollution Prevention Hearables Function

In today's world, where pollution poses a significant threat to the environment and our health, innovative technologies are emerging to combat this growing issue. One such groundbreaking development is pollution prevention hearables. These cutting-edge devices not only provide us with immersive audio experiences but also actively work towards reducing pollution. Let's delve into how pollution prevention hearables function and why they are a game-changer in the world of sustainable technology.

Pollution prevention hearables are designed to monitor and analyze the surrounding air quality in real-time. Equipped with advanced sensors, these sleek devices are capable of detecting harmful pollutants such as particulate matter, volatile organic compounds (VOCs), carbon monoxide, and more. By continuously monitoring the air around us, pollution prevention hearables offer valuable insights into the quality of the environment we live in.

Once pollution prevention hearables detect pollutants, they employ a range of innovative techniques to mitigate their impact. One of the primary methods involves the use of filters and purifiers integrated into the device. These filters trap and neutralize airborne contaminants, allowing users to breathe cleaner air. This not only benefits the wearer but also contributes to the overall reduction of pollution in the environment.

In addition to filtration, pollution prevention hearables also leverage advanced algorithms and artificial intelligence to provide personalized recommendations and alerts. These intelligent devices can learn from individual habits and patterns, adapting to the user's lifestyle and surroundings. For instance, if the air quality drops below a certain threshold, pollution prevention hearables can notify the user to take necessary precautions, such as wearing a mask or avoiding specific areas with high pollution levels.

Moreover, pollution prevention hearables are designed with sustainability in mind. They are built using eco-friendly materials and utilize energy-efficient components, reducing their environmental footprint. By promoting sustainable practices, these devices align with the growing trend of conscious consumerism and contribute to the niche of "hearable beauty."

Overall, pollution prevention hearables are revolutionizing the way we perceive technology and its impact on the environment. By combining the functionalities of immersive audio experiences and pollution reduction, these devices offer a unique and holistic approach to sustainability. As we move towards a greener future, pollution prevention hearables play a vital role in raising awareness, empowering individuals, and collectively working towards a cleaner and healthier planet.

Advantages and Limitations of Pollution Prevention Hearables

In recent years, the concept of pollution prevention hearables has gained significant attention and popularity, especially within the niche of hearable beauty. These innovative devices offer numerous advantages in terms of sustainability and personal well-being. However, like any technology, they also come with certain limitations. In this subchapter, we will explore the advantages and limitations of pollution prevention hearables.

Advantages:

1. Environmental Impact: Pollution prevention hearables play a crucial role in reducing environmental pollution. By actively monitoring and analyzing the surrounding air quality, these devices can help individuals make informed decisions about their environment, contributing to a cleaner and healthier planet.

2. Health Benefits: Air pollution is a major concern, especially in urban areas. Pollution prevention hearables can detect harmful pollutants such as particulate matter and alert the wearer to take necessary precautions. This not only protects the health of individuals but also promotes a sense of well-being.

3. Personalized Approach: Hearable beauty enthusiasts can enjoy the advantage of customization with pollution prevention hearables. These devices can be personalized to match individual preferences, whether it's through the choice of design, color, or additional features. This allows users to express their unique style while actively participating in environmental preservation.

4. Real-time Data: Pollution prevention hearables provide real-time data about air quality, allowing users to make immediate decisions

based on accurate information. Whether it's choosing a less polluted route or avoiding high-risk areas, these devices empower individuals to actively engage in pollution prevention.

Limitations:

1. Cost: One of the primary limitations of pollution prevention hearables is the cost associated with these devices. As they incorporate advanced technology and sensors, they tend to be more expensive than traditional hearables. This can make them inaccessible to some individuals, limiting their potential impact.

2. Accuracy: While pollution prevention hearables provide valuable data, their accuracy can vary. Factors such as sensor calibration, device placement, and environmental conditions can affect the reliability of the data provided. Users should be aware of this limitation and consider it when making decisions based on the information received.

3. Dependency on Connectivity: Pollution prevention hearables heavily rely on connectivity to provide real-time data and notifications. In areas with weak or no connectivity, the effectiveness of these devices may be compromised, limiting their functionality.

In conclusion, pollution prevention hearables offer numerous advantages in terms of environmental impact, health benefits, personalization, and real-time data. However, their limitations, such as cost, accuracy, and dependency on connectivity, should also be considered. As technology continues to advance, it is essential to address these limitations and strive for more accessible and reliable pollution prevention hearables. By doing so, we can truly harness the potential of these devices to create a sustainable and beautiful future for everyone.

Chapter 3: Types of Pollution Prevention Hearables

Noise-Canceling Headphones

In today's fast-paced world, where noise pollution has become an integral part of our daily lives, finding solace in quiet moments can be a challenging task. Fortunately, technology has come to our rescue with the invention of noise-canceling headphones. These innovative devices are designed to provide a sanctuary of tranquility amidst the chaos of the modern world.

Noise-canceling headphones are a true marvel of engineering. They work by actively reducing unwanted ambient sounds using advanced sound-cancellation technology. By analyzing the incoming sound waves, these headphones generate a mirror-image sound wave that cancels out the external noise, giving the listener a serene audio experience.

The benefits of noise-canceling headphones are numerous and extend beyond just blocking out noise. They allow individuals to immerse themselves in their favorite music, podcasts, or audiobooks without any distractions. Whether you're commuting to work, studying in a busy cafe, or simply looking for a moment of relaxation, noise-canceling headphones provide a personal sound bubble that elevates your listening experience to a whole new level.

Not only do noise-canceling headphones enhance our auditory experience, but they also have the potential to positively impact our mental health. Research has shown that excessive noise levels can lead to stress, anxiety, and even cognitive decline. By reducing external noise, these headphones can help create a calm and peaceful environment, promoting a sense of well-being and improving focus and productivity.

For those who value both functionality and style, the world of noise-canceling headphones offers a wide range of options. From sleek and minimalist designs to bold and vibrant colors, there is a headphone to suit every taste. Manufacturers have recognized the need to cater to the niche of "hearable beauty" by combining cutting-edge technology with aesthetically pleasing designs, ensuring that you not only enjoy an unparalleled audio experience but also make a fashion statement.

In conclusion, noise-canceling headphones are a game-changer in today's noisy world. They provide a sanctuary of tranquility, elevate our auditory experience, and promote mental well-being. With their ability to block out unwanted noise and their stylish designs, they have become an essential accessory for those seeking both functionality and style. So, whether you're a music lover, a student, or simply someone who craves moments of peace, noise-canceling headphones are a must-have item for everyone. Embrace the sounds of sustainability and immerse yourself in a world of pure audio bliss.

Air Quality Monitoring Earbuds

Air Quality Monitoring Earbuds: A Revolutionary Approach to Environmental Awareness

In today's fast-paced world, where personal well-being and sustainability go hand in hand, there is a growing need for innovative solutions that seamlessly integrate into our daily lives. Enter Air Quality Monitoring Earbuds, a groundbreaking technology that combines the realms of hearable beauty and environmental consciousness.

These cutting-edge earbuds not only deliver immersive sound and enhance your auditory experience, but they also serve as your personal air quality monitor, empowering you to take charge of your surroundings like never before. With these earbuds, you become a guardian of your own well-being and a champion for a cleaner, greener planet.

Imagine strolling through a bustling city street, the rhythmic beats of your favorite music resonating in your ears, while simultaneously being aware of the air quality around you. These earbuds discreetly capture data on pollutants, allergens, and toxins present in the air, providing you with real-time updates on the air quality index. Armed with this information, you can make informed decisions about where to go, when to wear a mask, or even when to take a breather in a nearby park.

But it doesn't stop there. Air Quality Monitoring Earbuds take this concept a step further by connecting to your smartphone via a user-friendly app. This app not only displays the air quality data collected by your earbuds but also offers personalized recommendations to help you mitigate potential health risks. It suggests alternative routes with better air quality, local green spaces to recharge in, or even alerts you to upcoming pollution hotspots.

The convergence of hearable beauty and environmental awareness brings forth a new era of sustainability. Air Quality Monitoring Earbuds not only provide you with a seamless auditory experience but also empower you to make conscious choices as you navigate the world. By being aware of the air you breathe, you become an active participant in the fight against pollution, contributing to a healthier planet for yourself and future generations.

So, whether you're a music enthusiast seeking a heightened audio experience or a sustainability advocate looking to make a positive impact, these earbuds are designed for you. Embrace this revolutionary technology, and join the movement towards a cleaner, more sustainable future. With Air Quality Monitoring Earbuds, you have the power to listen to the world and protect it at the same time.

Water Pollution Detection Earpieces

In today's world, where environmental concerns are at the forefront of discussions, it is essential to find innovative ways to prevent and detect pollution. One such groundbreaking invention that combines both functionality and style is the Water Pollution Detection Earpieces. These hearables not only enhance your auditory experience but also serve as an ingenious tool to help protect our precious water resources.

Imagine walking along a serene riverbank, listening to the soothing sounds of nature, when suddenly your earpieces alert you to the presence of water pollution. With Water Pollution Detection Earpieces, you can effortlessly identify and report pollutants, ensuring the well-being of our aquatic ecosystems.

These earpieces are designed to detect various contaminants, including chemicals, heavy metals, and harmful microorganisms, in bodies of water such as rivers, lakes, and oceans. Equipped with state-of-the-art sensors and advanced technology, these earpieces analyze the composition of water in real-time, allowing you to become proactive in protecting our environment.

The beauty of Water Pollution Detection Earpieces lies not only in their functionality but also in their aesthetic appeal. Created with the concept of "hearable beauty" in mind, these earpieces are stylish, comfortable, and customizable to suit your personal taste. Whether you prefer a sleek and minimalist design or a vibrant and eye-catching look, these hearables are a fashion statement in their own right.

By incorporating these earpieces into your everyday life, you become an active participant in the fight against water pollution. The earpieces connect seamlessly with your smartphone, providing you with an easy-to-use interface to monitor and track pollution levels. You can receive instant notifications, access pollution maps, and even contribute to a global database of water quality information, making a significant impact on the preservation of our planet's water resources.

Water Pollution Detection Earpieces are not just a gadget; they are a tool for change. They empower individuals to take responsibility for protecting our environment and inspire a collective effort towards sustainability. Every person can make a difference, and with these innovative hearables, you can contribute to a cleaner and healthier world for future generations.

In conclusion, the Water Pollution Detection Earpieces are a revolutionary invention that combines style and functionality to address the crucial issue of water pollution. By adopting these hearables, you can actively participate in the preservation of our planet's water resources, all while looking fashionable. Let us embrace the concept of "hearable beauty" and work together towards a sustainable and pollution-free future.

Soil Contamination Prevention Devices

In our pursuit of sustainable living, it is crucial to address the issue of soil contamination. The health of our soil directly impacts the quality of the food we consume and the overall well-being of our planet. To combat this problem, innovative technologies and devices have emerged, focusing on soil contamination prevention. These devices not only contribute to a healthier environment but also promote the concept of hearable beauty, merging aesthetics with sustainability.

One such device is the SoilGuard, a compact and user-friendly tool that detects soil contamination levels. Designed to be worn as a fashionable hearable, it alerts users to potential contamination hotspots in their vicinity. Equipped with advanced sensors, the SoilGuard can identify harmful chemicals and pollutants, providing real-time data on soil quality. This information empowers individuals to make informed decisions when it comes to gardening, farming, or even choosing the location for their next outdoor activity.

Another remarkable device is the SoilShield, a wearable hearable that acts as a protective shield against soil contamination. This device, resembling an elegant bracelet or pendant, utilizes cutting-edge nanotechnology to create a barrier between the skin and the soil. By preventing direct contact, it reduces the risk of harmful substances entering the body through the skin. The SoilShield not only safeguards our health but also adds a touch of style to our everyday lives, catering to the niche of hearable beauty.

Furthermore, the SoilSense is an innovative device that measures soil moisture and nutrients. This hearable, available in various designs and colors, provides real-time data on the condition of the soil. By monitoring these factors, individuals can optimize their water and fertilizer usage, ensuring optimal plant growth while minimizing the risk of contaminating the soil with excess chemicals. The SoilSense promotes sustainable gardening practices and adds a touch of eco-consciousness to the concept of hearable beauty.

In conclusion, soil contamination prevention devices offer a unique blend of sustainability and style. These hearables not only protect our health but also enable us to make environmentally conscious choices. By integrating technology, innovation, and fashion, these devices cater to the niche of hearable beauty, appealing to a wide audience. Let us embrace these devices and take a step towards a more sustainable and beautiful future.

Chapter 4: Applications of Pollution Prevention Hearables

Noise Pollution Reduction

Noise pollution is a growing concern in our modern society. With the constant hustle and bustle of urban environments, it can feel impossible to escape the relentless noise that surrounds us. From traffic and construction to loud music and noisy neighbors, excessive noise not only disturbs our peace of mind but also poses serious health risks.

In this subchapter, we will explore the concept of noise pollution reduction and its significance in creating a sustainable and harmonious environment. We will delve into the ways in which noise pollution affects our daily lives, our physical and mental health, and the steps we can take to minimize its impact.

Noise pollution not only disrupts our tranquility but also affects our overall well-being. Constant exposure to loud noises can lead to stress, anxiety, and even hearing loss. Additionally, it can disturb our sleep patterns, leading to fatigue and decreased productivity. Therefore, it is essential to address this issue and find effective solutions for noise pollution reduction.

Fortunately, advancements in technology have paved the way for innovative solutions in the form of hearables. These devices, which combine functionality and style, offer a unique opportunity to combat noise pollution while enhancing our auditory experience. This subchapter will explore the concept of "hearable beauty," where these devices not only serve a practical purpose but also add an element of aesthetic appeal to our daily lives.

We will discuss the various types of hearables available in the market, such as noise-canceling headphones and earbuds, which help us create our personal sound sanctuary even in the noisiest of environments. Moreover, we will explore the concept of customized hearables, designed to fit individual preferences and seamlessly integrate into our daily routines.

Furthermore, we will delve into the role of hearables in promoting sustainability. By reducing noise pollution, we can create a more sustainable environment that benefits both our physical and mental health. We will discuss the positive impact of noise pollution reduction on wildlife, the preservation of natural habitats, and the overall well-being of communities.

In conclusion, this subchapter on noise pollution reduction aims to raise awareness about the detrimental effects of excessive noise and the importance of finding practical solutions. By embracing the concept of "hearable beauty," we can not only enhance our auditory experience but also contribute to a more sustainable and harmonious world. Whether you are a tech-savvy individual or someone passionate about preserving the beauty of our natural surroundings, this subchapter will provide valuable insights to inspire you to take action against noise pollution.

Air Pollution Monitoring and Protection

In today's world, environmental issues have become a growing concern for everyone. One such critical issue is air pollution, which affects our health and well-being. To combat this problem, innovative technologies have been developed to monitor and protect against air pollution. This subchapter explores the concept of air pollution monitoring and protection, with a focus on the emerging field of "hearable beauty."

Air pollution monitoring involves the use of advanced sensors and devices to measure the quality of the air we breathe. These sensors can detect various pollutants such as particulate matter, nitrogen dioxide, and volatile organic compounds. By monitoring air pollution levels, individuals and communities can make informed decisions about their activities and take necessary precautions to minimize exposure.

One fascinating aspect of air pollution monitoring is the integration of this technology into hearables – wearable devices that enhance our auditory experience. Hearables, which include headphones, earbuds, and other wearable audio devices, have gained immense popularity due to their convenience and functionality. However, the potential of hearables goes beyond just delivering sound.

Enter the world of "hearable beauty," where these devices not only provide an immersive audio experience but also help protect us from air pollution. Hearables equipped with air pollution sensors can continuously monitor the air quality around us. They can provide real-time data on pollutant levels, alerting users to potential risks and prompting them to take necessary precautions.

Imagine going for a run in the park while your earbuds monitor the air quality and warn you when pollution levels exceed safe limits. Or enjoying a concert while your headphones analyze the surrounding air and provide recommendations on the best spots for fresh air breaks. These hearables not only enhance our auditory experience but also prioritize our health and well-being in polluted environments.

Furthermore, the data collected from these hearables can be aggregated and analyzed at a larger scale, providing valuable insights into air pollution patterns and trends. This information can be used to develop targeted pollution prevention strategies, improve urban planning, and create healthier living environments for everyone.

In conclusion, air pollution monitoring and protection have become crucial in our quest for sustainability. With the integration of sensors into hearables, we can now have a better understanding of the air quality around us and take necessary steps to protect ourselves. The concept of "hearable beauty" represents a remarkable fusion of technology and environmental consciousness, where our audio devices not only provide us with an immersive experience but also serve as guardians of our health in polluted surroundings. As we move forward, let us embrace these innovative solutions and work towards a cleaner and healthier future for all.

Water Pollution Prevention and Detection

Water pollution is a significant environmental issue that affects us all. It poses a threat to aquatic ecosystems, human health, and the overall sustainability of our planet. In this subchapter, we will explore the importance of water pollution prevention and detection, focusing on the innovative approach of "Sounds of Sustainability: Pollution Prevention Hearables."

Water pollution prevention begins with understanding the sources and causes of contamination. Industrial activities, improper waste disposal, agricultural runoff, and urban development are some of the leading contributors to water pollution. By raising awareness about these sources, we can encourage individuals, communities, and industries to adopt sustainable practices that minimize harmful effects on our water bodies.

One effective way to prevent water pollution is through the use of hearable beauty technology. The "Sounds of Sustainability: Pollution Prevention Hearables" is a groundbreaking innovation that combines fashion and functionality to help individuals actively contribute to water pollution prevention.

These hearables are stylish devices that can be worn as jewelry or integrated into clothing. Their primary function is to alert the wearer about potential sources of water pollution in their surroundings. Equipped with advanced sensors, these devices can detect chemical pollutants, excessive nutrient levels, and even harmful bacteria in the water.

The hearables emit subtle sounds or vibrations to indicate the presence of pollution. This unique approach allows individuals to be constantly aware of the water quality around them, empowering them to take immediate action. Whether it's reporting the pollution to local authorities or adjusting personal habits, these hearables encourage everyone to play an active role in preventing water pollution.

Furthermore, the hearables also serve as educational tools, providing information about the impact of pollution on ecosystems and human health. By raising awareness and fostering a sense of responsibility, this technology helps to create a culture of environmental stewardship.

The "Sounds of Sustainability: Pollution Prevention Hearables" not only contribute to water pollution prevention but also promote a sense of personal style and individual expression. By integrating fashion and functionality, these devices make sustainability accessible and attractive to a wider audience.

In conclusion, water pollution prevention and detection are crucial for maintaining the health and sustainability of our water bodies. Through the innovative approach of "Sounds of Sustainability: Pollution Prevention Hearables," individuals can actively contribute to preventing water pollution while expressing their personal style. By leveraging technology and raising awareness, we can create a world where water pollution is a thing of the past.

Soil Contamination Prevention and Rehabilitation

In the world of environmental sustainability, soil contamination prevention and rehabilitation play a crucial role. Our soil is not only the foundation for plant growth but also the source of our food, clean water, and overall ecological balance. Therefore, it becomes imperative for everyone to understand the significance of soil conservation and take necessary steps to protect it from contamination.

Soil contamination occurs when harmful substances, such as heavy metals, pesticides, or industrial waste, infiltrate the soil matrix. These contaminants can have detrimental effects on both human health and the environment. However, by implementing preventive measures, we can mitigate the risks associated with soil contamination.

One of the key approaches to soil contamination prevention is sustainable agriculture practices. Farmers and gardeners can adopt organic farming methods, which eliminate the use of harmful chemicals and promote natural soil fertility. By using compost, crop rotation, and biological pest control methods, we can nourish the soil while minimizing the risk of contamination.

Another important aspect is proper waste management. Industrial waste, hazardous materials, and household pollutants should be disposed of responsibly, ensuring they do not seep into the soil. Recycling and reusing materials, reducing the use of single-use plastics, and encouraging responsible waste disposal practices can significantly reduce the chances of soil contamination.

In addition to prevention, soil rehabilitation is equally critical. Remediation techniques are employed to restore contaminated soils back to their natural state, making them safe for agricultural use and promoting biodiversity. These techniques include soil vapor extraction, bioremediation, and phytoremediation, which use plants to remove contaminants from the soil.

To raise awareness about soil contamination prevention and rehabilitation, the concept of "hearable beauty" comes into play. Hearables, such as headphones or earbuds, have become a popular accessory in today's society. By integrating sustainability-focused content into the hearable experience, individuals can learn about soil conservation while enjoying their favorite tunes or podcasts.

Imagine listening to an uplifting playlist that not only entertains but also educates about the importance of soil health. Or tuning in to a podcast that explores success stories of soil rehabilitation projects and the positive impact they have on communities and ecosystems. By incorporating sustainability-focused content into hearables, we can engage a wider audience and inspire them to take action in preserving our precious soil.

In conclusion, soil contamination prevention and rehabilitation are vital for the sustainability of our planet. By implementing sustainable agriculture practices, proper waste management, and employing remediation techniques, we can protect our soil from contamination and restore already contaminated areas. Integrating sustainability-focused content into hearables presents a unique opportunity to raise awareness about soil conservation among a diverse audience. Let us all take responsibility and work collectively to ensure a healthy and thriving soil for generations to come.

Chapter 5: Technological Innovations in Pollution Prevention Hearables

Artificial Intelligence Integration

In the fast-paced world of technology, the integration of artificial intelligence (AI) has become increasingly prominent in various industries. From healthcare to transportation and beyond, AI has made its mark in revolutionizing the way we live and work. Now, it is making its way into the realm of hearable beauty, bringing with it a whole new level of sophistication and convenience.

The integration of AI in hearables is an exciting development that promises to enhance our auditory experiences while ensuring sustainability. These innovative devices are not just about delivering high-quality sound but are also designed to minimize pollution and promote a more eco-friendly lifestyle.

With AI integration, hearables can adapt and learn from the user's preferences, making adjustments to provide personalized sound experiences. Imagine a pair of earbuds that can analyze your hearing patterns and automatically adjust the audio levels to suit your unique needs. Whether you are listening to music, taking a call, or even enjoying a podcast, AI can optimize the sound quality to give you an immersive and enjoyable experience.

Furthermore, AI can also play a crucial role in pollution prevention. By monitoring environmental factors such as noise levels and air quality, hearables can provide real-time data to help users make informed decisions. For instance, AI-powered hearables can alert users when they are in a noisy or polluted environment, encouraging them to take necessary precautions or relocate to a quieter and cleaner area.

The integration of AI in hearables also opens up exciting possibilities for sustainable living. These devices can track and analyze personal energy consumption, making users more aware of their carbon footprint. By providing recommendations and reminders, AI can encourage individuals to adopt energy-saving practices, thereby contributing to a greener and more sustainable future.

Moreover, AI integration can enable seamless connectivity and communication. With built-in virtual assistants, hearables can respond to voice commands, allowing users to access information, manage tasks, and stay connected without the need for handheld devices. This hands-free approach not only enhances convenience but also promotes safety by minimizing distractions.

The integration of artificial intelligence in hearable beauty is a significant step towards a more sustainable and technologically advanced future. By combining personalized sound experiences, pollution prevention features, and sustainable living practices, these AI-powered devices have the potential to transform the way we perceive and interact with our auditory world. Embrace the sounds of sustainability and embark on a journey towards a greener, more connected, and immersive future.

Internet of Things Connectivity

In this rapidly advancing digital era, the Internet of Things (IoT) has become an integral part of our lives. From smart homes to wearable devices, IoT connectivity has revolutionized the way we interact with technology. This subchapter will delve into the concept of IoT connectivity and its significance in the context of "Sounds of Sustainability: Pollution Prevention Hearables."

IoT connectivity refers to the ability of devices to connect and communicate with each other over the internet. This connectivity enables the exchange of data and information, allowing devices to work together seamlessly. In the context of our pollution prevention hearables, IoT connectivity plays a crucial role in enhancing their functionality and effectiveness.

By harnessing the power of IoT, our pollution prevention hearables can seamlessly connect to various smart devices, such as smartphones, air quality monitors, and other wearable technology. This connectivity enables real-time data collection and analysis, providing users with accurate and up-to-date information about their surrounding environment.

For instance, imagine wearing a pollution prevention hearable that not only monitors the air quality around you but also communicates with your smartphone. The hearable can collect data on pollutants, noise levels, and other environmental factors, and then transmit this information to your smartphone. You can then receive notifications, alerts, and personalized recommendations based on this data, empowering you to make informed decisions to protect your health and the environment.

Moreover, IoT connectivity allows for remote control and customization of our pollution prevention hearables. Through dedicated mobile applications, users can adjust settings, track their pollution exposure history, and even contribute their data to larger environmental databases. This collective data can then be analyzed to identify pollution patterns, support research, and drive policy changes for a cleaner and healthier world.

The possibilities of IoT connectivity in the realm of "hearable beauty" are endless. By seamlessly integrating our pollution prevention hearables with other smart devices and applications, users can enjoy a personalized and immersive experience. Whether it's receiving real-time pollution alerts, tracking personal exposure levels, or contributing to a global movement for sustainability, the IoT connectivity of our hearables fosters a sense of empowerment and collective action.

In conclusion, IoT connectivity is a pivotal aspect of our pollution prevention hearables. By connecting our devices to the internet and enabling seamless communication with other smart devices, we can harness the power of data to create a cleaner and healthier future. This subchapter has explored the significance of IoT connectivity in the context of "Sounds of Sustainability: Pollution Prevention Hearables," illustrating its potential to revolutionize the way we perceive and address environmental challenges.

Advanced Sensor Technologies

In today's fast-paced world, where technology plays an integral role in our daily lives, there is a growing need for innovative solutions that address the challenges of sustainability. One such area of development is advanced sensor technologies, which have the potential to revolutionize the way we perceive and prevent pollution. This subchapter explores the exciting realm of advanced sensor technologies and their application in pollution prevention hearables, catering to the niche of "hearable beauty."

Advanced sensor technologies refer to the cutting-edge techniques used to develop sensors that are highly sensitive, accurate, and versatile. These sensors can detect various environmental factors such as air quality, noise levels, and pollutants, all at a granular level. In the context of pollution prevention hearables, these technologies enable the creation of smart devices that not only provide aesthetic appeal but also actively contribute to a sustainable environment.

The integration of advanced sensor technologies into hearable beauty products has far-reaching implications. These devices can monitor air quality in real-time, alerting users to potential pollutants and encouraging them to take preventive measures. For instance, imagine wearing a stylish earpiece that not only enhances your appearance but also notifies you when you enter an area with high pollution levels, prompting you to wear a mask or find an alternative route. By leveraging advanced sensor technologies, such devices can significantly contribute to reducing our exposure to harmful pollutants and improving overall well-being.

Moreover, advanced sensor technologies can also enable personalized experiences in the realm of hearable beauty. By analyzing individual data, these sensors can provide tailored recommendations on skincare routines, suggesting suitable products based on the user's specific needs. Imagine a hearable device that not only plays your favorite music but also provides real-time updates on your skin's moisture levels, recommending the appropriate skincare regimen to maintain a healthy complexion.

In conclusion, the integration of advanced sensor technologies into pollution prevention hearables represents a significant leap towards sustainable living. These technologies offer a unique blend of aesthetics and functionality, ensuring that users not only look good but also actively contribute to a pollution-free environment. It is an exciting time for the hearable beauty niche, as advanced sensor technologies pave the way for a sustainable and personalized future. By embracing these advancements, we can all play a part in creating a cleaner, healthier, and more beautiful world for everyone.

Sustainable Materials in Hearable Devices

In recent years, the rise of hearable devices has revolutionized the way we experience sound and connect with technology. From wireless earbuds to smart headphones, these devices have become an integral part of our daily lives, enhancing our listening experiences and providing us with convenient features. However, with the growing concern for environmental sustainability, it is crucial to explore the use of sustainable materials in the production of these hearable devices.

The concept of sustainable materials refers to materials that are sourced, manufactured, and disposed of in an environmentally responsible manner. By incorporating sustainable materials into the design and manufacturing process of hearables, we can minimize the negative impact on the environment and promote a greener future.

One of the key aspects of sustainable materials in hearable devices is the choice of biodegradable or recyclable materials. Traditional earbuds and headphones are often made from plastic, which takes hundreds of years to decompose. By opting for biodegradable alternatives, such as bio-plastics made from renewable resources like cornstarch or bamboo, we can significantly reduce the long-term environmental impact of these devices.

Additionally, the use of recycled materials in hearable devices can greatly contribute to sustainability efforts. Materials like recycled aluminum, stainless steel, or even reclaimed ocean plastic can be utilized in the production process, reducing the need for virgin materials and minimizing waste generation.

Another aspect to consider is the energy efficiency of hearable devices. By incorporating energy-efficient components and optimizing power consumption, we can reduce the overall environmental footprint of these devices. This can be achieved through the use of low-power Bluetooth technology, efficient battery management systems, and renewable energy sources for charging.

Furthermore, sustainability can be promoted through the design and manufacturing processes themselves. Adopting modular designs that allow for easy repair and upgradeability can prolong the lifespan of hearable devices, reducing electronic waste generation. Additionally, implementing sustainable manufacturing practices, such as reducing water and energy consumption and minimizing the use of harmful chemicals, can further enhance the sustainability of these devices.

In conclusion, sustainable materials in hearable devices play a crucial role in mitigating the environmental impact of these innovative technologies. By choosing biodegradable or recyclable materials, incorporating recycled materials, optimizing energy efficiency, and adopting sustainable design and manufacturing practices, we can ensure a greener and more sustainable future for the hearable beauty industry. As consumers, it is essential to support brands that prioritize sustainability, as our choices can make a significant difference in preserving the beauty of our planet for generations to come.

Chapter 6: Challenges and Opportunities in Environmental Protection Hearables

Legal and Regulatory Considerations

In the fast-paced world of technology and innovation, it is crucial to understand the legal and regulatory landscape surrounding hearable beauty products. As the demand for these cutting-edge devices continues to rise, it is essential to ensure that they comply with all relevant laws and regulations to protect consumers and promote sustainability. This subchapter explores the key legal and regulatory considerations that must be taken into account when developing and using hearable beauty devices.

One of the primary legal considerations is intellectual property rights. As the field of hearable beauty evolves, companies must be mindful of patent protection to safeguard their innovations from being copied or stolen. It is important to conduct thorough research and file for patents to secure exclusive rights to their unique designs and functionalities. Additionally, trademarks should be obtained to protect the brand identity and prevent confusion among consumers.

Another critical aspect to consider is product safety regulations. Hearable beauty devices, like any other electronic product, must meet certain safety standards to ensure they do not pose any harm to users. Compliance with applicable regulations, such as the Federal Communications Commission (FCC) guidelines in the United States, is crucial to guarantee that the devices are safe for use and do not emit harmful levels of radiation.

Moreover, environmental regulations play a significant role in promoting sustainability in the production and disposal of hearable beauty products. Companies must adhere to waste management regulations and ensure proper disposal of electronic components to prevent environmental pollution. Materials used in hearable beauty devices should be carefully selected, taking into account their environmental impact and recyclability.

Data protection and privacy are also important considerations in the development of hearable beauty products. As these devices collect and analyze personal data, companies must adhere to data protection laws to ensure the privacy and security of their users' information. Compliance with regulations such as the General Data Protection Regulation (GDPR) in the European Union is crucial to maintain user trust and confidence.

In conclusion, legal and regulatory considerations are of utmost importance in the field of hearable beauty. By understanding and complying with intellectual property rights, product safety regulations, environmental standards, and data protection laws, companies can develop sustainable and legally compliant hearable beauty products. This ensures the protection of consumers and promotes innovation in this exciting and rapidly growing niche.

Ethical Implications of Hearable Technologies

In this subchapter, we will explore the ethical implications surrounding the use of hearable technologies, specifically in the context of "hearable beauty." As these innovative devices become more prevalent in our daily lives, it is crucial to critically examine the potential ethical concerns that arise.

One of the primary ethical concerns surrounding hearable technologies is privacy. These devices are designed to seamlessly integrate into our lives, constantly capturing and processing data about our habits, preferences, and even biometric information. While this data can be used to enhance our experiences and provide personalized services, it also raises concerns about the security and ownership of our personal information. Companies must prioritize data protection and ensure that users have control over their data to mitigate potential privacy breaches.

Another ethical consideration is the potential for addiction and dependence on hearable technologies. As these devices become more advanced and immersive, individuals may find it challenging to disconnect from them, leading to an overreliance on technology. This dependence can have adverse effects on mental health and interpersonal relationships. It is crucial for individuals to strike a balance between using hearable technologies for their benefits and maintaining a healthy detachment from them.

Furthermore, the environmental impact of hearable technologies should not be overlooked. The production and disposal of these devices contribute to electronic waste, which poses significant environmental challenges. Manufacturers should adopt sustainable practices, prioritize recycling, and design devices with longevity in mind to reduce their ecological footprint.

Additionally, the ethical implications of hearable technologies extend to the potential for discrimination and inequality. These devices have the power to gather vast amounts of data, which, if used irresponsibly, can perpetuate biases and discrimination in various sectors such as healthcare, employment, and finance. Developers and policymakers must implement safeguards to prevent the misuse of data and ensure equitable access to hearable technologies for all individuals.

In conclusion, while hearable technologies offer numerous benefits, it is essential to acknowledge and address the ethical implications that arise from their use. Privacy concerns, addiction, environmental impact, and potential discrimination are just a few of the ethical considerations that require careful consideration. By actively engaging in discussions surrounding these issues, individuals, manufacturers, and policymakers can work together to ensure the ethical and sustainable development and use of hearable technologies in the niche of hearable beauty.

Economic and Market Trends

In the ever-evolving world of sustainability, it is crucial to keep abreast of the latest economic and market trends. Understanding these trends not only helps us make informed decisions about our purchasing habits, but also empowers us to contribute to a better future for our planet. This subchapter explores the various economic and market trends related to hearable beauty, ensuring that we are well-equipped to make sustainable choices.

One significant trend that has emerged in recent years is the growing consumer demand for eco-friendly and sustainable products. People are becoming increasingly conscious of the environmental impact of their purchases and are actively seeking out alternatives that align with their values. Hearable beauty, which combines the benefits of wearable technology with sustainable materials and production methods, has gained substantial traction in this regard.

Another trend that cannot be overlooked is the rise of the circular economy. This economic model aims to minimize waste and maximize the resources we already have. In the context of hearable beauty, this means designing products that can be repaired, recycled, or repurposed, rather than being disposed of after a short lifespan. By embracing the principles of the circular economy, hearable beauty brands can not only reduce their environmental footprint but also foster a more sustainable market.

Furthermore, the market trends in hearable beauty have witnessed a remarkable shift towards transparency and accountability. Consumers now demand more information about the materials used, the manufacturing processes employed, and the overall sustainability practices of the brands they support. This increased transparency allows consumers to make well-informed choices and hold companies accountable for their environmental impact.

Additionally, innovation and technological advancements play a crucial role in shaping the economic and market trends of hearable beauty. As new materials are developed, such as biodegradable plastics and sustainable textiles, the possibilities for creating eco-friendly hearable beauty products expand. Moreover, emerging technologies like 3D printing can revolutionize the manufacturing process, enabling customization and reducing waste.

In conclusion, understanding the economic and market trends is essential for anyone interested in the niche of hearable beauty. By being aware of the growing demand for sustainable products, the principles of the circular economy, the need for transparency, and the role of innovation, we can make choices that align with our values while contributing to a more sustainable future.

Future Opportunities and Potential Development

In the fast-paced world we live in, technology is constantly evolving and reshaping the way we live and interact with our environment. One such area of technological advancement is hearables, which have gained immense popularity in recent years. These innovative devices are not only changing the way we listen to music or take phone calls, but they also hold great potential in addressing various environmental concerns. This subchapter explores the future opportunities and potential development of hearables, specifically in the niche of "hearable beauty."

Hearables have the unique ability to enhance our sensory experiences, allowing us to immerse ourselves in a world of sounds and melodies. However, their potential goes beyond just entertainment. Imagine a future where you can wear a pair of earbuds that not only deliver high-quality audio but also serve as pollution prevention devices. These hearables could be equipped with advanced sensors that detect harmful pollutants in the air, alerting the wearer and providing real-time data on air quality. This would enable individuals to make informed decisions about their environment and take necessary actions to protect their health.

Furthermore, the concept of "hearable beauty" opens up a whole new realm of possibilities. Imagine customized earbuds that not only fit perfectly but also incorporate sustainable materials and designs. These hearables could be made from biodegradable or recycled materials, reducing waste and contributing to a more sustainable future. Additionally, they could be adorned with intricate patterns and vibrant colors, transforming them into fashionable accessories that enhance your personal style.

The potential development of "hearable beauty" extends to the realm of self-expression. Imagine being able to personalize the sounds you hear based on your mood or preferences. Hearables could incorporate advanced sound adjustment features, allowing you to create a unique and immersive sonic experience tailored to your liking. This would not only revolutionize the way we listen to music but also provide a new avenue for artistic expression and creativity.

As we look towards the future, it is evident that hearables have immense potential to address environmental concerns and revolutionize the way we experience sound. The concept of "hearable beauty" brings together sustainability, fashion, and personalization, offering individuals a new way to express themselves while contributing to a cleaner, greener world. The possibilities are endless, and it is up to us to embrace this technology and unlock its full potential.

In conclusion, the future opportunities and potential development of hearables, particularly in the niche of "hearable beauty," hold great promise. The integration of advanced sensors, sustainable materials, and personalized sound experiences opens up a world of possibilities for individuals to protect the environment, express themselves, and enhance their sensory experiences. By embracing these advancements, we can pave the way for a more sustainable and harmonious future.

Chapter 7: Case Studies on Pollution Prevention Hearables

Case Study 1: Noise Reduction Hearables in Urban Environments

Introduction:

Urban environments are known for their bustling streets, constant traffic, and never-ending construction projects. While cities offer a multitude of opportunities, they also come with a significant downside - noise pollution. Excessive noise can lead to stress, sleep disturbances, and even long-term health issues. To combat this problem, innovative technology has emerged in the form of noise reduction hearables. These devices not only provide a solution to noise pollution but also contribute to the niche of "hearable beauty," enhancing our auditory experience while promoting sustainable living.

The   Problem:
Living in a city can be overwhelming, especially for those seeking moments of peace and tranquility. The constant drone of traffic, honking horns, and chattering crowds can be mentally and physically draining. Traditional noise-cancelling headphones have been an option, but they often come with bulky designs that hinder portability and aesthetics. This is where noise reduction hearables step in, offering a stylish and efficient solution.

The   Solution:
Noise reduction hearables are sleek devices designed to counteract the negative effects of noise pollution while providing an immersive and enjoyable auditory experience. These hearables utilize cutting-edge technology, such as active noise cancellation algorithms and advanced

sound processing, to filter out unwanted noise and deliver crystal-clear audio.

How Noise Reduction Hearables Work: These devices use tiny microphones to detect external noises and generate opposing sound waves to cancel them out. By actively reducing ambient noise, users can enjoy their chosen audio content without distraction. Whether it's music, podcasts, or audiobooks, noise reduction hearables ensure that the user's auditory experience remains uninterrupted and enjoyable.

The Beauty of Hearables: Hearables have evolved beyond their functionality; they have become a fashion statement. Manufacturers have recognized the need for stylish designs that seamlessly integrate into our daily lives. Noise reduction hearables are available in various shapes, sizes, and colors, allowing users to express their individuality while enjoying the benefits of noise reduction technology. These sleek devices not only enhance our auditory experience but also add a touch of elegance to our appearance.

Conclusion:
Noise reduction hearables have revolutionized the way we experience sound in urban environments. They offer a practical and aesthetically pleasing solution to noise pollution, allowing users to enjoy a peaceful soundscape amidst the chaos of city life. By embracing the niche of "hearable beauty," these devices contribute to sustainable living by promoting the importance of noise reduction and fostering a sense of personal style. With noise reduction hearables, everyone can transform their urban auditory experience into one of tranquility and beauty.

Case Study 2: Air Quality Monitoring Hearables in Industrial Settings

In the pursuit of sustainable solutions, the integration of technology and innovation has become vital. One such groundbreaking development is the advent of air quality monitoring hearables in industrial settings. These devices not only contribute to pollution prevention but also serve as a testament to the beauty of wearable technology.

Industrial activities often result in the release of pollutants, which pose significant health risks and environmental threats. Understanding the importance of addressing this issue, researchers and engineers have collaborated to create hearables capable of monitoring air quality in real-time.

Imagine a scenario where workers in a bustling industrial setting can effortlessly keep track of their immediate environment's air quality. These hearables, designed to be sleek and aesthetically pleasing, blend seamlessly into the concept of hearable beauty. They can be worn discreetly, resembling fashionable earpieces or jewelry, allowing individuals to monitor air quality without drawing unnecessary attention.

This case study delves into the successful implementation of air quality monitoring hearables in an industrial setting. By equipping workers with these devices, employers can ensure a safe and healthy working environment. Real-time data collected by these hearables allows for quick and informed decision-making, enabling prompt action to mitigate pollution sources.

Furthermore, these hearables go beyond the individual level. Data collected from various devices can be aggregated and analyzed to identify pollution hotspots, monitor emission trends, and develop effective pollution prevention strategies. This collective effort plays a pivotal role in creating a sustainable and pollution-free future.

The integration of air quality monitoring hearables in industrial settings not only enhances worker safety but also showcases the potential of wearable technology. It bridges the gap between functionality and fashion, proving that sustainability and style can coexist harmoniously.

In conclusion, this case study highlights the revolutionary concept of air quality monitoring hearables in industrial settings. By addressing pollution prevention and emphasizing the beauty of wearable technology, these devices provide a glimpse into a sustainable future. From ensuring worker safety to facilitating data-driven decision-making, these hearables contribute to a cleaner and healthier environment. As we embrace the possibilities of hearable beauty, we move one step closer to a more sustainable and pollution-free world.

Case Study 3: Water Pollution Detection Hearables in Marine Environments

Introduction:

In this case study, we explore the revolutionary concept of Water Pollution Detection Hearables in Marine Environments. These innovative devices not only enhance our understanding of pollution in our oceans but also bring together the realms of technology and environmental conservation. By wearing these hearables, individuals can actively contribute to protecting our marine ecosystems while embracing the concept of "hearable beauty."

Understanding Water Pollution in Marine Environments: Water pollution poses a significant threat to marine life and ecosystems. It is crucial to monitor and detect pollution levels accurately to ensure timely actions are taken. Traditional methods of data collection can be time-consuming and costly. This is where Water Pollution Detection Hearables come into play.

How do Water Pollution Detection Hearables Work? Water Pollution Detection Hearables are wearable devices equipped with advanced sensors that detect and analyze various pollutants in marine environments. These devices are designed to be compact, efficient, and aesthetically pleasing, catering to the niche of "hearable beauty." By wearing these devices, individuals can contribute to real-time data collection efforts without compromising their style or comfort.

The Role of Water Pollution Detection Hearables: The primary purpose of these hearables is to empower individuals to actively participate in environmental conservation. By wearing these devices while engaging in water-based activities, such as swimming, snorkeling, or boating, users can collect valuable data on water quality. This data can be shared with relevant authorities, researchers, and organizations working towards pollution prevention in marine ecosystems.

Benefits and Impact:
The integration of technology and environmentalism through Water Pollution Detection Hearables has several advantages. Firstly, it enables individuals to develop a personal connection with their surroundings and actively engage in pollution prevention. Secondly, the data collected from these devices can contribute to large-scale efforts in monitoring pollution levels, identifying sources, and implementing effective solutions. Finally, the concept of "hearable beauty" ensures that users can express their personal style while making a positive impact on the environment.

Conclusion:
Water Pollution Detection Hearables in Marine Environments offer a unique and promising solution to combat water pollution and protect our precious marine ecosystems. By embracing the concept of "hearable beauty," individuals can actively contribute to pollution prevention efforts while expressing their personal style. These devices empower everyone to become environmental stewards and play an active role in ensuring the sustainability of our oceans.

Case Study 4: Soil Contamination Prevention Hearables in Agricultural Practices

In recent years, the detrimental effects of soil contamination on agricultural practices have become increasingly apparent. The excessive use of chemical fertilizers and pesticides has led to a significant decline in soil quality, posing a threat to both human health and the environment. However, advancements in technology have opened new doors for sustainable farming practices. One such innovation is the development of soil contamination prevention hearables, which have revolutionized the way we approach agriculture.

These hearables, designed to be worn by farmers during their work, provide real-time data and analysis on soil quality. Equipped with advanced sensors, they are capable of measuring various parameters such as pH levels, nutrient content, and the presence of harmful chemicals. By constantly monitoring these factors, farmers can take immediate action to prevent or reduce soil contamination, ultimately leading to healthier crops and a more sustainable farming ecosystem.

The integration of hearables in agricultural practices has proven to be highly effective in preventing soil contamination. By alerting farmers to potential issues, such as excessive pesticide use or nutrient imbalances, these devices allow for timely interventions. For instance, if the pH level of the soil is found to be too acidic, farmers can adjust their irrigation or add organic amendments to restore the balance. Similarly, if excessive levels of pesticides are detected, farmers can modify their spraying techniques or explore alternative pest control methods, reducing the risk of contamination.

Furthermore, these hearables promote precision farming, optimizing resource utilization and minimizing environmental impact. By providing accurate data on soil conditions, farmers can tailor their irrigation, fertilizer application, and crop rotation practices accordingly. This targeted approach not only reduces the use of chemicals but also enhances crop productivity, leading to improved economic outcomes for farmers.

The implementation of soil contamination prevention hearables in agricultural practices is a significant step towards sustainable and environmentally friendly farming. By empowering farmers with real-time information and promoting precision farming techniques, these devices contribute to the preservation of soil quality and the reduction of chemical inputs. Ultimately, their adoption can lead to a more resilient agricultural sector, capable of meeting the challenges of feeding a growing global population without compromising the health of our planet.

In conclusion, the integration of soil contamination prevention hearables in agricultural practices represents a remarkable advancement in sustainable farming. These devices enable farmers to monitor soil quality in real-time and make informed decisions to prevent contamination. By promoting precision farming techniques and reducing the use of chemicals, hearables contribute to the long-term health of our soils and the preservation of our environment. With their potential to transform agricultural practices, these hearables are revolutionizing the way we approach farming and highlighting the importance of sustainable solutions in our quest for a greener future.

Chapter 8: The Future of Pollution Prevention Hearables

Emerging Technologies and Innovations

In today's rapidly evolving world, emerging technologies and innovations are reshaping our lives in ways we could have never imagined. From the internet to artificial intelligence, these advancements have revolutionized various industries, including the field of "hearable beauty."

Hearables, a term used to describe wearable devices that enhance our auditory experience, have gained immense popularity in recent years. These devices not only provide us with high-quality sound but also offer a range of innovative features that cater to our aesthetic preferences. This subchapter explores the exciting world of hearable beauty, where technology and fashion converge to create a unique sensory experience.

One of the most remarkable innovations in hearable beauty is pollution prevention technology. With increasing concerns about air pollution and its detrimental effects on our health, hearables have emerged as a solution to mitigate these risks. These devices come equipped with advanced sensors that can detect pollutants in the surrounding environment. Through real-time monitoring, they provide wearers with valuable insights about air quality, enabling them to make informed decisions about their well-being.

Furthermore, hearable beauty is not limited to pollution prevention alone. These devices also incorporate cutting-edge design elements that cater to individual style preferences. With customizable designs, wearers can express their fashion sense while enjoying the benefits of these technologically advanced devices.

In addition to pollution prevention and aesthetics, hearable beauty also encompasses features that enhance our overall auditory experience. With noise-canceling technology, wearers can immerse themselves in their favorite music or podcasts without any external distractions. Moreover, the integration of augmented reality capabilities allows users to enhance their daily activities, from workouts to meditation, by providing a personalized audio experience.

As these technologies continue to evolve, the potential for hearable beauty to contribute to a sustainable future becomes even more exciting. Imagine a world where pollution is actively monitored and reduced through wearable devices that seamlessly blend with our personal style. This subchapter delves into the possibilities of this emerging field and explores the potential for hearable beauty to transform our lives in ways that are both aesthetically pleasing and environmentally conscious.

In conclusion, emerging technologies and innovations have paved the way for the exciting field of hearable beauty. With pollution prevention technology, customizable designs, and enhanced auditory experiences, these devices offer a unique blend of fashion and technology. As we embrace these advancements, we are not only enhancing our personal style but also contributing to a more sustainable future. The possibilities are limitless, and the journey towards a harmonious fusion of technology and aesthetics has only just begun.

Integrating Hearables with Sustainable Lifestyles

In today's fast-paced world, the concept of sustainability is gaining momentum, and individuals are becoming more conscious of their impact on the environment. As we strive to adopt sustainable practices in various aspects of our lives, it is important to consider how technology can contribute to this cause. One such technology that has the potential to revolutionize the way we live sustainably is hearables.

Hearables, a term used to describe wearable devices for audio listening and communication, have gained popularity in recent years. These devices, including wireless earbuds, headphones, and smart hearing aids, have become an integral part of our daily lives. They enhance our audio experiences and provide us with convenience and connectivity. But what if these devices could also contribute to our efforts towards a sustainable lifestyle?

The concept of integrating hearables with sustainable lifestyles is an exciting prospect. Imagine a world where your wireless earbuds not only deliver exceptional sound quality but also actively help in reducing pollution. This subchapter explores the potential of incorporating sustainable features into hearable devices, focusing specifically on the niche of hearable beauty.

Hearable beauty refers to the aesthetics and design of hearable devices. By infusing eco-friendly materials, such as biodegradable plastics or sustainable wood, into the construction of these devices, manufacturers can reduce their carbon footprint. Additionally, advancements in solar-powered technology can enable hearables to harness energy from the sun, reducing the need for disposable batteries and promoting renewable energy sources.

Furthermore, hearables can be integrated with pollution monitoring capabilities. Imagine wearing earbuds that not only deliver your favorite tunes but also provide real-time information about the air quality around you. By analyzing pollutants in the environment, these devices can alert users to high pollution levels, encouraging them to take necessary precautions or even influence lifestyle changes to reduce their impact on the environment.

Integrating hearables with sustainable lifestyles is not just about reducing pollution; it's also about empowering individuals to make informed decisions. By incorporating features like audio-guided sustainable living tips or connecting users with eco-friendly communities, hearables can inspire and educate individuals on how to lead a more sustainable life.

In conclusion, the integration of hearables with sustainable lifestyles holds immense potential. By infusing sustainable materials, utilizing renewable energy sources, and incorporating pollution monitoring capabilities, hearables can become powerful tools in our journey towards a greener, more sustainable future. With hearable beauty as a niche, we can enhance our audio experiences while making a positive impact on the environment. It's time to embrace technology as a catalyst for change and create a harmonious balance between innovation and sustainability.

Potential Impact on Environmental Conservation Efforts

In our quest for sustainable living and protecting the environment, it is crucial to explore every avenue that can contribute to conservation efforts. One such avenue that holds immense potential is the field of hearable beauty. These innovative devices not only enhance our listening experience but also have the power to make a positive impact on environmental conservation.

Hearables, which encompass a wide range of audio devices like headphones, earphones, and smart hearing aids, have gained popularity due to their ability to provide high-quality sound and personalized audio experiences. However, their potential impact on environmental conservation efforts is often overlooked. By examining this aspect, we can unlock new possibilities for a greener future.

One significant way hearables can contribute to environmental conservation is through the use of sustainable materials. Many manufacturers are now incorporating eco-friendly materials such as bio-plastics, recycled metals, and organic fabrics in their products. By choosing these sustainable materials, we can reduce the carbon footprint and minimize the environmental harm caused by traditional manufacturing processes.

Furthermore, hearables can play a crucial role in raising awareness about environmental issues. These devices can be equipped with built-in sensors that detect noise pollution levels, air quality, and even measure carbon dioxide levels. By providing real-time data and personalized feedback, hearables can empower individuals to make informed decisions that positively impact the environment. People can adjust their daily routines, choose green transportation options, or participate in community initiatives to reduce pollution and conserve resources.

In addition to awareness, hearables can also foster a deeper connection with nature. Imagine being able to listen to the sounds of a rainforest or a babbling brook while wearing your headphones. By experiencing the beauty of nature through sound, hearables can inspire a sense of awe and appreciation for our natural world. This emotional connection can motivate individuals to take action in preserving and protecting the environment.

To maximize the potential impact of hearables on environmental conservation efforts, it is crucial for manufacturers, consumers, and policymakers to work together. Manufacturers can continue to develop sustainable and eco-friendly products, while consumers can choose to support these initiatives by investing in environmentally responsible hearables. Policymakers can also play a role by implementing regulations that encourage the use of sustainable materials and promote awareness campaigns.

In conclusion, the field of hearable beauty offers immense potential for environmental conservation efforts. By utilizing sustainable materials, raising awareness, and fostering a deeper connection with nature, hearables can become powerful tools in our mission to protect the environment. Let us embrace this technology and work together to create a greener and more sustainable future for all.

Chapter 9: Conclusion

Recap of Key Points

In this subchapter, we will recap the key points discussed throughout the book "Sounds of Sustainability: Pollution Prevention Hearables," addressing an audience of everyone interested in the niches of hearable beauty. We have explored the fascinating world of pollution prevention hearables and how they contribute to sustainable living and personal well-being.

First and foremost, we emphasized the importance of understanding the impact of pollution on our environment and health. Pollution is a global issue that affects everyone, and it is crucial that we take proactive measures to mitigate its effects. Pollution prevention hearables offer a unique solution by combining technology, fashion, and environmental consciousness.

We delved into the concept of hearable beauty, which refers to the integration of aesthetics and functionality in pollution prevention hearables. These devices not only protect us from harmful pollutants but also enhance our appearance, allowing us to express our individual style. By embracing hearable beauty, we can promote sustainable fashion choices while safeguarding our health.

Throughout the book, we highlighted various types of pollution prevention hearables available in the market. From air purifying earrings and necklaces to pollution monitoring smartwatches, these devices utilize cutting-edge technology to detect and filter pollutants in our surroundings. By wearing them, we can create personal microenvironments that are free from harmful substances.

We also discussed the importance of integrating sustainable materials and production methods into the manufacturing of hearables. By using eco-friendly materials and implementing ethical practices, we can reduce our ecological footprint and contribute to a more sustainable future. Additionally, we explored the concept of upcycling, where old hearables can be repurposed into new designs, reducing waste and promoting circularity.

Lastly, we emphasized the role of education and awareness in fostering a sustainable mindset. By spreading knowledge about pollution prevention and the benefits of hearables, we can inspire individuals to make informed choices and actively participate in creating a cleaner and healthier environment.

In conclusion, pollution prevention hearables offer a creative and effective solution to combat pollution while embracing personal style. By understanding the key points discussed in this book, we can contribute to a sustainable future and foster a sense of hearable beauty. Let us embrace these innovative devices, protect our environment, and promote a healthier lifestyle for ourselves and future generations.

Final Thoughts on Pollution Prevention Hearables

In today's fast-paced world, pollution has become a significant concern that affects not only our environment but also our health. As we strive to find innovative solutions to combat pollution, one area that has gained attention is hearable technology. Pollution Prevention Hearables have emerged as a revolutionary concept, combining the benefits of wearable devices with a focus on environmental sustainability. This subchapter delves into the final thoughts on Pollution Prevention Hearables, exploring their potential impact on both the general audience and the niche of hearable beauty.

For everyone, Pollution Prevention Hearables offer a unique opportunity to play an active role in preserving our planet. These wearable devices are equipped with cutting-edge technology that allows users to monitor and reduce their exposure to harmful pollutants. By providing real-time data on air quality, noise pollution, and other environmental factors, hearables empower individuals to make informed decisions about their surroundings. Whether you are a student, a working professional, or a concerned citizen, Pollution Prevention Hearables can enable you to take proactive steps towards a healthier and more sustainable lifestyle.

Moreover, the niche of hearable beauty can greatly benefit from Pollution Prevention Hearables. As individuals become increasingly conscious of the impact of pollution on their appearance, these innovative devices offer a comprehensive solution. Pollution Prevention Hearables not only monitor the levels of pollutants in the environment but also provide personalized recommendations for skincare and beauty routines. By seamlessly integrating technology and beauty, hearables can help individuals maintain healthy and radiant skin, even in polluted environments.

In conclusion, Pollution Prevention Hearables are a promising avenue for combating pollution and promoting sustainability. By combining technology, environmental awareness, and personal well-being, these devices have the potential to revolutionize the way we interact with our environment. Whether you are an individual concerned about pollution or a beauty enthusiast seeking innovative solutions, Pollution Prevention Hearables offer a holistic approach towards a cleaner and healthier future. Embracing this technology can empower each one of us to make a positive impact on the environment while enhancing our own well-being. Let us embrace this opportunity and contribute to a sustainable world through Pollution Prevention Hearables.

Call to Action for a Sustainable Future

In a world where the environment is facing unprecedented challenges, it is imperative that we all take action to secure a sustainable future for generations to come. The concept of "hearable beauty" has emerged as a powerful tool in promoting environmental consciousness and pollution prevention. This subchapter will delve into the importance of this concept and provide a call to action for individuals from all walks of life.

Hearable beauty is a term coined to describe the fusion of technology and sustainability in wearable devices that are both aesthetically pleasing and environmentally friendly. These hearables not only deliver cutting-edge audio experiences but also serve as a reminder of our responsibility towards the planet. By embracing this concept, we can integrate sustainable practices into our daily lives while enjoying the benefits of technological advancements.

The call to action begins with each and every one of us. Regardless of our background or niche interests, we must recognize the urgent need to prioritize sustainability. Whether you are a fashion enthusiast, a fitness lover, or a tech-savvy individual, there are countless ways to contribute to a sustainable future through your choices.

To start, let's consider the materials used in the production of hearable beauty products. Opting for recycled or upcycled materials reduces the demand for virgin resources while minimizing waste. By supporting brands that prioritize sustainability in their manufacturing processes, we can drive positive change in the industry.

Additionally, we should encourage the development of renewable energy sources to power these hearables. Embracing solar or kinetic energy technology not only reduces our carbon footprint but also fosters innovation in the field. Imagine a world where our devices are powered by the very movements we make or the sunlight that surrounds us.

Another crucial aspect of the call to action is the responsible disposal of electronic waste. Instead of discarding old or broken devices, we should explore recycling programs and take advantage of initiatives that promote proper e-waste management. By doing so, we prevent harmful substances from leaching into the environment while extracting valuable materials for future use.

Ultimately, the call to action for a sustainable future is a collective effort. It is not enough for a select few to champion this cause; we must all actively participate in creating a world that is both beautiful and sustainable. By embracing hearable beauty and integrating sustainable practices into our daily lives, we can pave the way for a brighter future. Let us join hands, take action, and make a positive impact on the world we inhabit. The time for change is now.